Orlando Austin New York San Diego Toronto London

Visit *The Learning Site!*
www.harcourtschool.com

Position and Motion

Are you near a window right now? Are you a few feet away from a friend? Are you sitting to the right or left of your teacher? Everything has a position. **Position** is the location of an object.

Sometimes a position doesn't change. For instance, the position of your ears on the sides of your head stays the same. Sometimes a position does change, such as when you move from one room to another—or even when you just take a step. When an object changes position, it's called **motion**.

Popcorn explodes with motion!

This could be a frame of reference for you if the car is moved later.

Motion happens in all directions—forward, backward, up, down, straight, and curving. Think of a roller coaster or a bag of popcorn in a microwave. The roller coaster moves forward, up, and down. It also goes straight and curves along a track. Popcorn moves all around inside the bag when it's being popped.

Let's think about how we understand the movement of objects. Suppose you look at two photographs taken on the same day. In one photo, a car is parked in front of your house, far away from your driveway. In the other, the same car is closer to your driveway. You know that the car has moved because you know where the driveway is—it is your frame of reference. You compared the position of the driveway to the position of the car. A frame of reference is something stationary that you compare movement to.

COMPARE AND CONTRAST How would you compare the idea of position to the idea of motion?

Fast and Slow

Words like fast and slow describe the speed of something. **Speed** is a way to tell how much an object's position has changed over time.

Objects that are fast, such as airplanes and rockets, change their position quickly. Objects that are slow, such as merry-go-rounds, change their position slowly.

The best way to understand speed is to use measurements. First, you need to know the distance, or how far the object has traveled. Second, you need to know the time it took to travel the distance. Then you can find the speed by dividing the distance by the time.

For example, Chicago, Illinois, and St. Louis, Missouri, are about 480 kilometers apart. If it takes you six hours to drive between the two cities, you travel at about 80 kilometers per hour.

A merry-go-round goes at a slow speed.

A roller coaster's speed changes as it goes up and down the track.

Here's something else to keep in mind: Not all objects move at the same speed all the time. Some machines, like elevators or fans, move at a steady speed. Others don't.

For instance, when we say a car is traveling at a speed of 80 kilometers (50 mi) per hour, we are taking an average. Sometimes the driver might be traveling at 85 kilometers (53 mi) per hour. At other times, when the car slows down for tolls or exits, the car could be traveling between 15 and 25 kilometers (9 and 16 mi) per hour.

COMPARE AND CONTRAST How can you contrast the steadiness in speed of a slowly moving motorboat with that of a moving rowboat?

Describing Speed

When we think about how fast an object is traveling, we don't usually think only about its speed. We also think of the direction it's going. For example, if you saw a car driving on a highway with a tire going flat, you might call the police for help and tell them that there's a "white station wagon going 55 miles an hour, heading north." You have just given information about the car's velocity. **Velocity** is a measurement of an object's speed and direction. The car's velocity is 55 miles per hour, north.

To describe directions, we use words that tell us about location—right, left, up, down, north, south, east, or west. If two objects are traveling in opposite directions, they may have the same speed, but they will have different velocities.

You read earlier that a car doesn't always travel at the same speed. Sometimes it speeds up, sometimes it slows down, and sometimes it stops.

STOP

A car has to slow down and stop for this sign! Stopping is also acceleration.

A change in direction is a change in velocity.

Cars also sometimes turn and change their direction. They can go backward when the driver puts the car into the reverse gear. These changes are called **acceleration**, or a change in the speed or direction of an object's motion.

Sometimes people think acceleration only has to do with speeding up. It also has to do with slowing down. Acceleration is any change in velocity.

CAUSE AND EFFECT What will happen to the velocity of a car that makes a right turn at a street corner?

Feeling the Force

When you were younger, an older person probably helped you ride a playground swing by pushing you. Any time something is pulled or pushed, that is **force**. When you were being pushed, the person was using force to move your body. Did the person also pull you way back to give your swinging a good start? Force was used then, too.

When the older person used force, you moved. Unless you used your own force to move the swing, if the other person didn't push you, you would have stayed in one place. Force changes motion.

Any change of speed or direction needs a force. Force causes acceleration.

The direction of the force changes the direction in which an object moves. If the older person pushed you to the side instead of forward, you would move to the side.

This swing doesn't go anywhere unless force is used.

A large push against the tennis ball will make it move a lot!

What would happen if one person stood in front of the swing, and the other person stood behind the stopped swing, and they both pushed with the same force? Nothing would happen! They are two forces of the same size going in opposite directions. They will cancel out each other.

The size of a force also affects acceleration. A small push against a swing will make it move a little. A big push will make the swing move a lot.

CAUSE AND EFFECT **What happens to an object when force is used on it?**

Force is measured in newtons (N). This word honors the work of Isaac Newton, who did early research involving force. Newton's Second Law helps us understand how forces make objects accelerate.

Mass and Force

How much force does it take to move an object? It depends on its mass. It takes more force to accelerate a large mass than it does to accelerate a small mass to the same speed.

Think of the difference between a fallen tree and a kitten. It would take very little energy to move a kitten. However, you might not want to move a tree without some help!

All matter has **inertia**—the property that keeps an object at rest or moving in a straight line. A ball that is rolling in one direction on a flat, straight surface tends to keep going in the same direction. Inertia also has to do with matter that is not moving. The fallen tree is sitting in one place, so it will take a force to get it moving.

It will take a lot of force to move this load of lumber.

The more mass there is in an object, the more inertia it has. It's harder to change its motion. If the tree is rolling in one direction, it will be very difficult to get it to change direction!

Sometimes we need a lot of force to move objects, so we use tools and machines. You could tie ropes to the tree and pull it, but it would still be difficult. You would probably use a machine, because you would need a greater force than your own muscles can give.

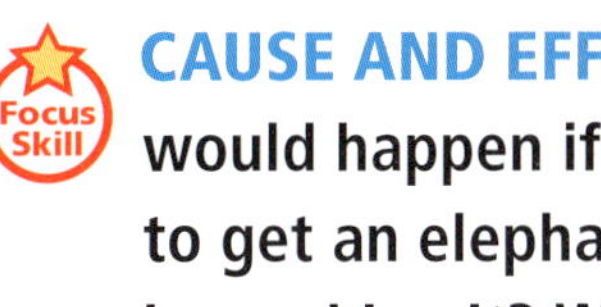

CAUSE AND EFFECT What would happen if you tried to get an elephant to move by pushing it? Why?

This elephant has a lot of inertia!

Gravity and Weight

We use force to move our bodies. We also use electric forces to help us run machines, and there are electric charges in nature.

Gravity is the force that keeps us on the ground by pulling us toward Earth. What happens when you throw a ball up in the air? It comes down!

Gravity is an example of gravitation—the force that causes all masses to attract one another. Gravitation keeps the moon orbiting around Earth. The moon and Earth are pulled toward each other.

Gravitation enables Earth to orbit the sun. Gravitation also keeps all of the other planets pulled toward the sun, as they move in their orbits.

Fast Fact

Gravity at the equator is slightly weaker than at the North Pole because Earth bulges at the equator and you are farther from the center of Earth.

The force of gravity pulls the planets and the sun toward one another.

Your weight on the moon would be different from your weight on Earth.

The larger and closer a mass is to another mass, the more it affects the object. We are affected by the gravity of Earth because Earth is very large and very close to us.

You can measure the force of gravity by measuring weight on a scale. **Weight** is the gravitational force acting on an object.

Remember that mass and weight are different. Your mass on the moon would be the same—you have the same amount of matter inside you wherever you are. Your weight would change because the gravitational force on the moon is different from that of Earth.

MAIN IDEA AND DETAILS How does gravity control the relationship between the planets and the sun?

Facts of Friction

In addition to force that moves objects, there is a force called friction that can slow down objects. **Friction** resists, or fights, motion between objects that are touching.

Anytime two objects are touching, there is friction between their surfaces. Have you ever rubbed your hands together to keep warm on a cold day? You were producing friction. You sensed that the friction would produce heat, and you were correct! Friction changes kinetic energy (motion) to heat.

When you've walked on a gymnasium floor, you've probably noticed a big difference if you've been wearing leather-soled shoes instead of sneakers. The sneakers have rubber soles that increase friction and keep you from slipping on the smooth wooden floor.

Anything we use that has brakes, such as cars and bicycles, uses friction. The brakes have rubber pads that press against a moving surface to cause friction.

Tires help us ride smoothly, but they have to generate some friction so we can stop when we need to.

Surfaces can wear out due to friction, and sometimes we want objects, such as gears, to move smoothly without friction. We can make surfaces smooth or add machine oil to reduce friction when we need to.

Think of a baseball player racing to get to a base. He doesn't want too much friction to stop him. He slides on the ground, using the smoothness of his baseball uniform and his body to keep him moving as quickly as possible.

Fast Fact

Tires have to be able to stop on wet roads. If a tire is too smooth, it is not safe to ride on. That's why car tires should be inspected for wear every 10,000–15,000 miles.

MAIN IDEA AND DETAILS Give an example of how friction works.

A smooth, waxed wood floor helps a bowling ball move quickly.

Summary

Position is where an object is located. Motion is the change of position of an object. Objects move in different directions and at different speeds, and we can measure their movement. Force also causes motion. The amount of force needed to move an object depends on the object's mass. There are forces at work all around us, including natural forces and gravity.

Glossary

acceleration (ak•sel•er•AY•shuhn) Any change in the speed or direction of an object's motion (7, 8, 9)

force (FAWRS) A pull or push of any kind (8, 9, 10, 11, 12, 13, 14, 15)

friction (FRIK•shuhn) A force that resists motion between objects that are in contact (14, 15)

gravity (GRAV•ih•tee) The force of attraction between Earth and other objects (12, 13, 15)

inertia (in•ER•shuh) The property of matter that keeps an object at rest or keeps it moving in a straight line (10, 11)

motion (MOH•shuhn) A change of position of an object (2, 3, 7, 8, 11, 14, 15)

position (puh•ZISH•uhn) The location of an object (2, 3, 4, 15)

speed (SPEED) The measure of an object's change in position during a unit of time (4, 5, 6, 7, 8, 15)

velocity (vuh•LAHS•uh•tee) The measure of the speed and direction of an object's motion (6, 7)

weight (WAYT) A measure of the gravitational force acting on an object (13)